ULTIMATE SUPERCARS

ASTON MARTIN DB 11

By Craig Ellenport

Kaleidoscope
Minneapolis, MN

The Quest for Discovery Never Ends

· ·

This edition first published in 2021 by Kaleidoscope Publishing, Inc.

No part of this publication may be reproduced in whole or in part without written permission of the publisher.

For information regarding permission, write to
Kaleidoscope Publishing, Inc.
6012 Blue Circle Drive
Minnetonka, MN 55343

Library of Congress Control Number
2020936272

ISBN
978-1-64519-258-9 (library bound)
978-1-64519-326-5 (ebook)

Text copyright © 2021 by Kaleidoscope Publishing, Inc. All-Star Sports, Bigfoot Books, and associated logos are trademarks and/or registered trademarks of Kaleidoscope Publishing, Inc.

Printed in the United States of America.

FIND ME IF YOU CAN!

Bigfoot lurks within one of the images in this book. It's up to you to find him!

TABLE OF CONTENTS

Chapter 1: Pride of England 4

Chapter 2: Pride of England 10

Chapter 3: Speed and Style 16

Chapter 4: Getting Noticed 24

Beyond the Book 28
Research Ninja 29
Further Resources 30
Glossary 31
Index 32
Photo Credits 32
About the Author 32

Chapter 1
Movie Star Car

Benny just saw the car of his dreams. He didn't see it on the highway. It was not parked on the street. He saw it in a movie!

Benny and his dad had seen the James Bond movie. *No Time to Die* is the 25th James Bond movie. James Bond is a British secret agent. He saves the world from bad guys.

James Bond always drives a fancy sports car. In *No Time to Die*, Bond drives the Aston Martin DB11.

Thirteen Bond movies have featured an Aston Martin. In 1964's *Goldfinger*, Bond drove the Aston Martin DB5. Bond helped make Aston Martins popular!

FUN FACT

Bond's famous Aston Martin DB5 sold in 2019 for $6.4 million!

PARTS OF AN
ASTON MARTIN DB11

Clamshell hood

V12 twin turbo

Aeroblade

Bond's Aston Martins always look cool and drive fast. In the movies, the Aston Martins come with crazy gadgets, too. One car can shoot missiles. Another can travel underwater. Still another can become invisible!

You won't find Bond's Aston Martin on your street, however.

Bonded aluminum body

Available winter wheel kit

Aston Martins don't have gadgets like that in real life. Still, they all look cool and drive very fast.

Thanks to the Bond movies, Aston Martins have become famous everywhere. The movies inspire people to buy the cars. They also help Aston Martin designers create new looks and styles.

Benny was excited as they left the theater. "Did you see when Bond made the car skid?

"And what about when he was in that car chase?"

Benny had a great idea.

"Can we get a DB11?" Benny asked his dad.

Benny's dad smiled. "That would be great," he said. "But a new Aston Martin DB11 costs more than

A carbon fiber shell protects the DB11's V12 twin turbo engine. It can put out more than 600 horsepower.

two hundred thousand dollars. That's a lot of money!"

Then Benny's dad had a great idea, too.

"Tell you what. Let's visit the dealer. We can take a DB11 for a test drive."

Benny couldn't wait to take that ride. He would be just like James Bond!

Chapter 2
Pride of England

Just like James Bond, Aston Martin is the pride of England. Both spy and car are from Great Britain.

Lionel Martin and Robert Bamford started Aston Martin in 1913. They built a racing car first. Martin drove the car in a race called the Aston Clinton Hill Climb. The carmakers combined the name of the hill with the name of the driver. That's how they got the name: Aston Martin!

In 1922, the Aston Martin made its debut in the French Grand Prix. That was the top level of racing at the time. Aston Martin cars were among the fastest in the world.

Martin and Bamford had success making racing cars, but that business was slow. Aston Martin needed to grow. David Brown made that happen.

FUN FACT
Grand Prix [PREE] means "biggest prize" in French.

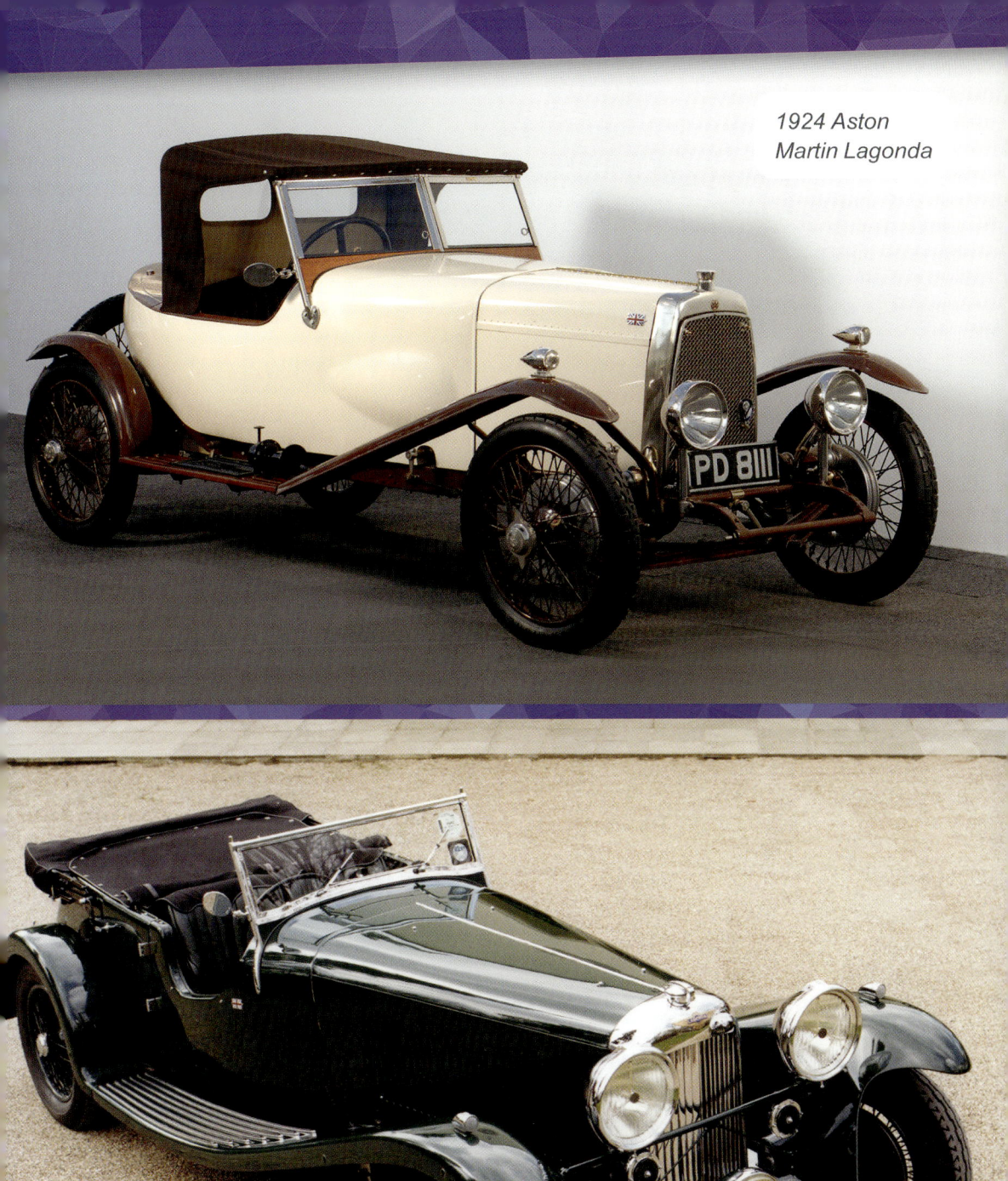

1924 Aston Martin Lagonda

1924 Aston Martin Bamford

David Brown bought Aston Martin Racing in 1947. Brown had a new plan for the company. He wanted Aston Martin to build more than just race cars. He wanted to combine fast performance with style.

Brown wanted the Aston Martin to look like a luxury car. He thought wealthy people would want to drive these fast, beautiful cars. He also wanted to make sure the inside was comfortable for both driver and passengers.

Aston Martin DB1

WHERE THE ASTON MARTIN DB11 IS MADE

Gaydon, Warwickshire, England

EGYPT TO ENGLAND

The Aston Martin logo looks like the wings of a bird. It's not. It is actually the wings of a scarab beetle.

The logo was designed by Sammy Davis. He was a driver for Aston Martin's racing teams. As an artist, Davis was interested in ancient Egypt. That culture looked on the scarab beetle as a symbol of great wealth and power.

The first Aston Martin model to roll out of the new factory was the Aston Martin DBR1. The "DB" in the name of the car stood for David Brown. The owner's initials would be used in many new Aston Martin models.

FUN FACT
Shelby later became a famous car designer for Ford.

In 1959, a DBR1 won the 24 Hours of LeMans. The 24 Hours of LeMans has been held every year in France since 1923. It is considered one of the most important car races in the world. Roy Salvadori of England and Carroll Shelby of the United States drove the car to victory. The race is so long that drivers have to take turns!

A restored 1957 DBR1

Chapter 3
Speed and Style

The DB11 is one of Aston Martin's newest models. It first came out in 2017. Every year since then, Aston Martin has made small changes to improve the driving performance and comfort of the DB11.

The DB11 comes in two styles. The **Coupe** has a hard top. The Volante is a **convertible**. The Volante's soft top opens wide. Both types combine a modern look with Aston Martin's classic design.

THE AEROBLADE

All fast cars are designed to be aerodynamic. Part of good aerodynamics is downforce. Downforce helps the tires grip the road better. That helps the car go faster. The force also makes turns smoother.

The DB11 creates downforce using the Aeroblade. This is a small opening in the front of the car. When the car is moving, air travels into the Aeroblade. It follows a path within the body of the car. That creates the downforce.

The DB11 comes with two types of motors. The twin turbo-charged V12 engine is in the DB11 AMR. That's the racing car. Look for this awesome machine on top race tracks in Europe. It is rare to see a V12 on local streets or highways.

The more common DB11 engine is the **V8**. That one is also twin turbo-charged. The V8 gets up to 503 horsepower. That is still much stronger than the

average car's engine. The popular Toyota Camry is only about 200 horsepower, for example.

The DB11 also lets the driver change how the engine behaves. They use GT mode for regular street driving. For longer trips, drivers can switch to Sport mode. For really fast driving, they can switch to Sport Plus.

A 2019 V8-powered DB11 Coupe

THE ASTON MARTIN IN DETAIL

Height: 4 feet, 2.4 inches (1.28 m)

Width: 6 feet, 9 inches (2.06 m)

LENGTH: 15 feet, 6.5 inches (4.72 m)

WEIGHT: 4,134 pounds (1,875 kg)

TOP SPEED: 208 miles per hour (334 kph)

TIME FROM 0-60 MPH: 3.7 seconds (96.6 kph)

COST: $200,000 (United States)

The inside of the DB11 is clean and comfortable. The two front seats and two back seats are all made of soft leather. New owners can also buy carbon fiber trim for decoration.

The dashboard has two video screens. There is a 12-inch screen in front of the driver that displays information about the car. In the middle, there is an 8-inch screen for music and other media.

FUN FACT
DB11 owners can buy luggage that matches their leather car seats.

Aston Martin likes its owners to feel special. They offer many different options for the interior of the car. The leather seats can come in different colors. The back of the seats can look like wood or metal. So can the insides of the doors.

The front seats can also be heated. This is a bonus for people driving in cold weather. A new feature for the 2020 DB11 is a heated steering wheel. That's a great feature, especially if you don't have gloves!

A heated steering wheel and lots of color choices: that's what is in store for DB11 owners. Also, some cars are made for British drivers. In that country, the driver sits on the right-hand side of the car.

Chapter 4
Getting Noticed

It was a few days after they saw the new James Bond movie when Benny and his dad visited an Aston Martin dealer.

Before they went, Benny read a story in *Road and Track* magazine. The writer was very impressed with the DB11. He wrote that people on the street would stop and stare when he drove by in the Aston Martin.

Benny read, "Nothing I've ever driven has swiveled heads like this DB11."

Benny agreed. He couldn't wait to see people staring at him and his dad. He would thrill people when he rode in the amazing Aston Martin DB11.

A NEW OWNER

Lawrence Stroll is a billionaire. He has his own Formula One racing team. In 2020, he became part-owner of Aston Martin. He paid about $213 million. Stroll's racing team will be known as Aston Martin F1 starting in 2021. Aston Martin fans will be able to root for the world's best drivers.

"I, and my partners, firmly believe that Aston Martin is one of the great global luxury car brands," said Stroll.

The day was warm and sunny. Benny and his dad walked into the Aston Martin showroom. They asked to test drive the DB11 Volante. Benny was excited to go for a ride with the top down!

Benny climbed into the passenger seat. He strapped himself in. Then his dad pulled out of the parking lot. The first thing Benny noticed was how smoothly the car turned.

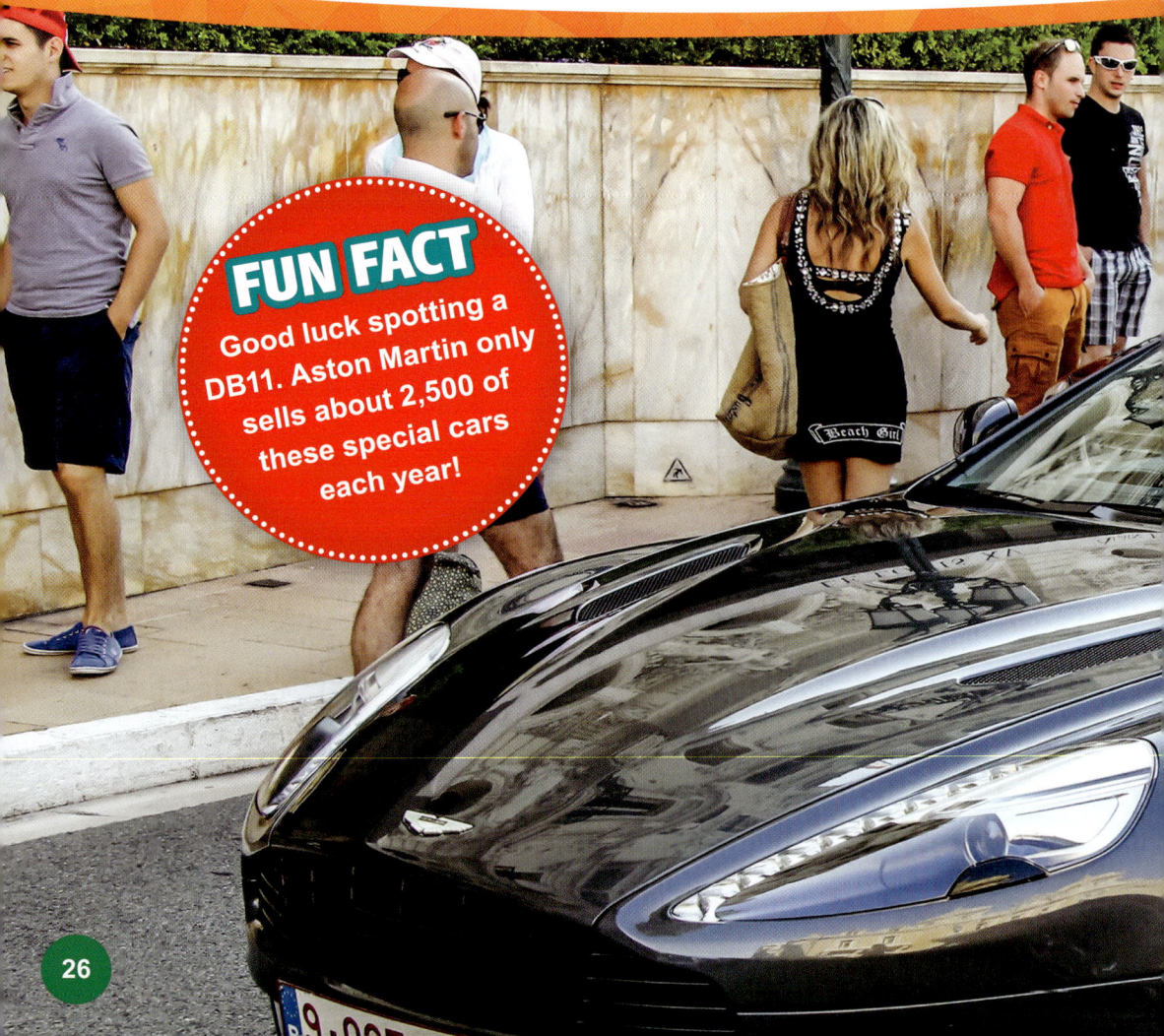

FUN FACT
Good luck spotting a DB11. Aston Martin only sells about 2,500 of these special cars each year!

Benny wasn't the only person impressed with the Aston Martin DB11. He could see people walking on the sidewalk. They were all staring at the DB11. Did they think it was James Bond?

No, it was not Bond this time. Instead, people simply recognized the very special Aston Martin DB11.

BEYOND THE BOOK

After reading the book, it's time to think about what you learned. Try the following exercises to jumpstart your ideas.

RESEARCH

FIND OUT MORE. Where would you go to find out more about your favorite cars? Find out what company makes the car and locate its website. What information do the companies provide? What other sources of car information can you find?

CREATE

GET ARTISTIC. Cars start with creative artists and designers. Time for you to take a shot! Get art materials and create a great, new car. Will you make it a sports car? A sedan? A race car? What colors will you paint it? What features can you give it? Let your imagination go for a spin!

DISCOVER

DIG DEEPER. James Bond movies made Aston Martins famous. What other famous movie cars can you think of? For fun, pick your favorite movie and design a car the heroes could use. What features would it have? What would it look like?

GROW

GO TO A CAR SHOW. Car shows are a great way to see lots of cool cars up-close. Check your local events calendar, or ask at a car dealer for upcoming events. You can find shows of old cars and new cars, sports cars and classic cars. Go to a show and find a new favorite car to love!

RESEARCH NINJA

Visit *www.ninjaresearcher.com/2589* to learn how to take your research skills and book report writing to the next level!

RESEARCH
DIGITAL LITERACY TOOLS

SEARCH LIKE A PRO
Learn about how to use search engines to find useful websites.

FACT OR FAKE?
Discover how you can tell a trusted website from an untrustworthy resource.

TEXT DETECTIVE
Explore how to zero in on the information you need most.

SHOW YOUR WORK
Research responsibly— learn how to cite sources.

WRITE

GET TO THE POINT
Learn how to express your main ideas.

PLAN OF ATTACK
Learn prewriting exercises and create an outline.

DOWNLOADABLE REPORT FORMS

Further Resources

BOOKS

Bond Cars and Vehicles. New York: DK Publishing, 2015.

Murray, Julie. *Aston Martin DB9.* Minneapolis: Abdo Publishing, 2018.

Noakes, Andrew. *Aston Martin DB: 70 Years.* London: White Lion Publishing, 2019.

WEBSITES

Factsurfer.com gives you a safe, fun way to find more information.

1. Go to www.factsurfer.com.
2. Enter "Aston Martin DB11" into the search box and click 🔍
3. Select your book cover to see a list of related websites.

Glossary

aerodynamic: having a shape that can move through air quickly.

convertible: a car with a roof that opens.

coupe: a car with a hard roof and two doors.

downforce: air pushing down on the car as it drives.

Formula One: an international, open-wheel racing circuit that features some of the world's fastest cars.

horsepower: a measurement of the power of an engine or motor to do work.

V8: an engine with 8 cylinders mounted on the crankshaft.

V12: an engine with 12 cylinders mounted on the crankshaft.

Index

24 Hours of Le Mans, 15
aeroblade, 6, 17
Aston Clinton Hill Climb, 10
Aston Martin DB5, 4
Aston Martin DBR1, 14
Bamford, Robert, 10
Bond, James, 4, 5, 6, 7, 8, 9, 10, 24, 27
Brown, David, 10, 12, 14
Davis, Sammy, 13
Egypt, 13
engine 8, 18, 19
England, 10, 13, 15
French Grand Prix, 10
Goldfinger, 4
interior, 22
Martin, Lionel, 10
No Time to Die, 4
Road and Track, 24
Salvadori, Roy, 15
Shelby, Carroll, 14, 15
Stroll, Laurence, 25
twin turbo engine, 6, 8, 18
Volante, 16, 26

PHOTO CREDITS

The images in this book are reproduced through the courtesy of: 123RF.com: Artem Konovalov 26. Courtesy Aston Martin: 6, 8, 10 (2), 12, 14, 16, 18, 19, 20, 23, 24. Shutterstock: Jeff Bukowski 4; Media Works 21; Auto-Data-Net 22.
Cover: Maksim Toome/Shutterstock (car); gyn9037/Shutterstock (background, top); zhao jiankang/Shutterstock (background, bottom).

About the Author

Craig Ellenport is an award-winning journalist, author, and editor from Massapequa, New York. He has authored numerous books and articles on subjects ranging from football and tennis to social media and health.